CONSTRUCTION TOOLS COLORING BOOK

This book
belongs to

Test color page

1. Screwdriver

A screwdriver is used to insert and remove screws

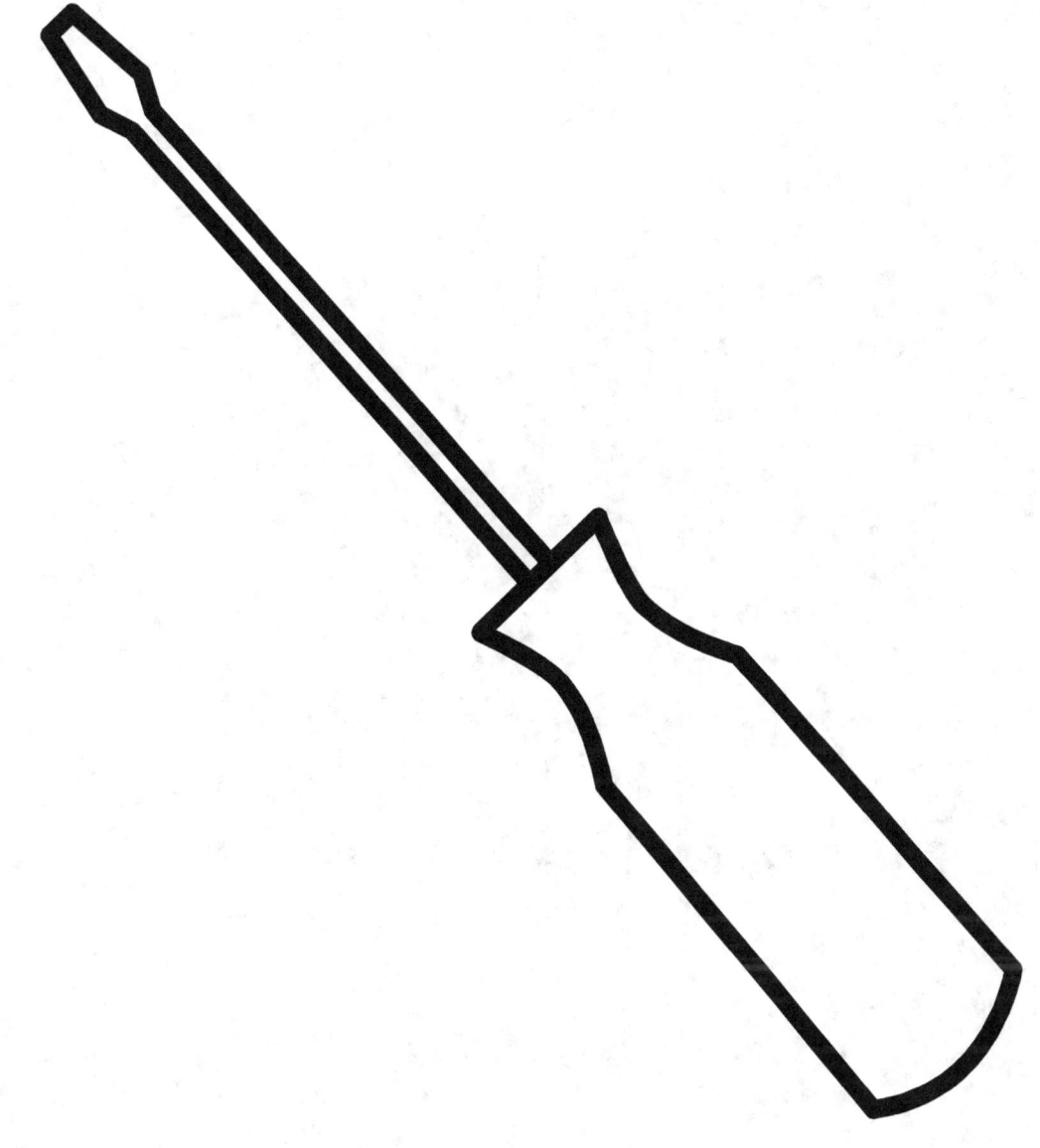

SCREWDRIVER

2. Screw

Screws are used to connect individual building elements to each other

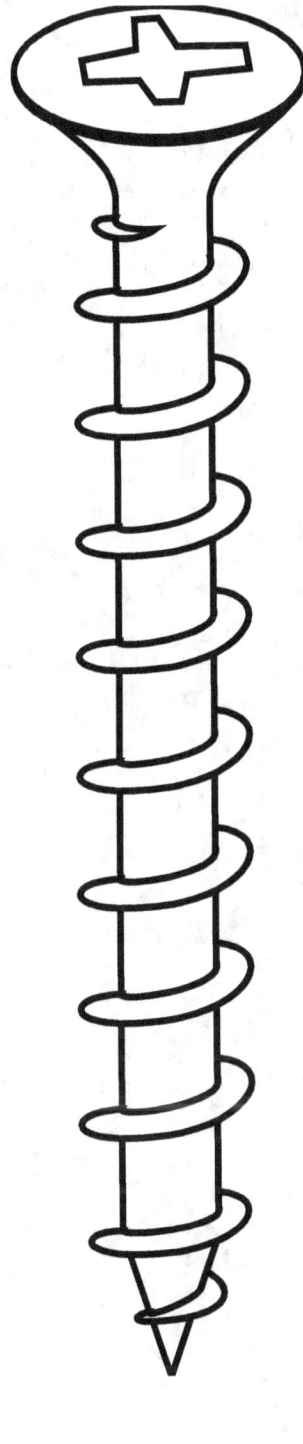

SCREW

3. Hammer

The hammer is used to drive nails

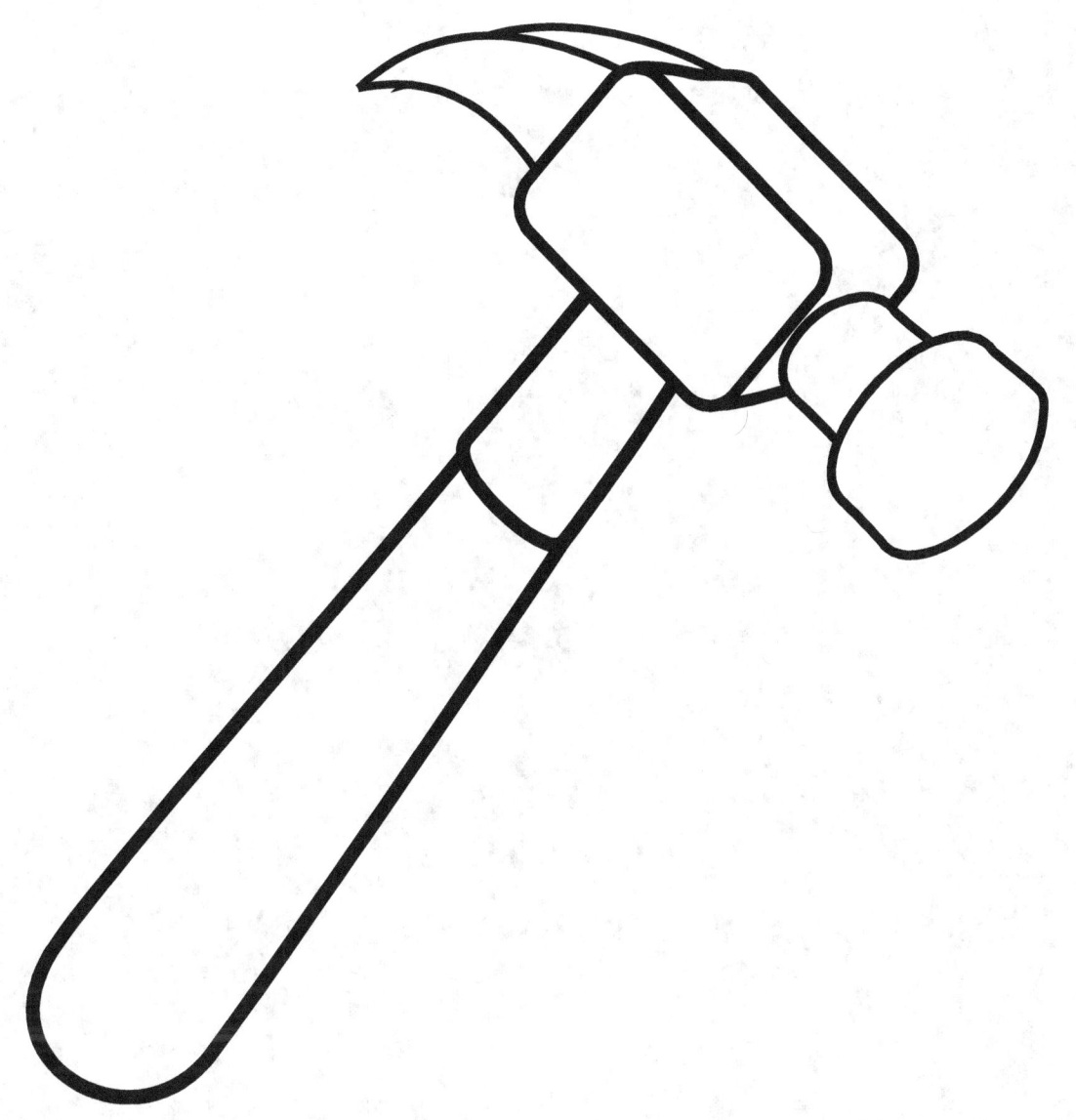

HAMMER

4. Nails

Nails are used to connect individual building elements to each other

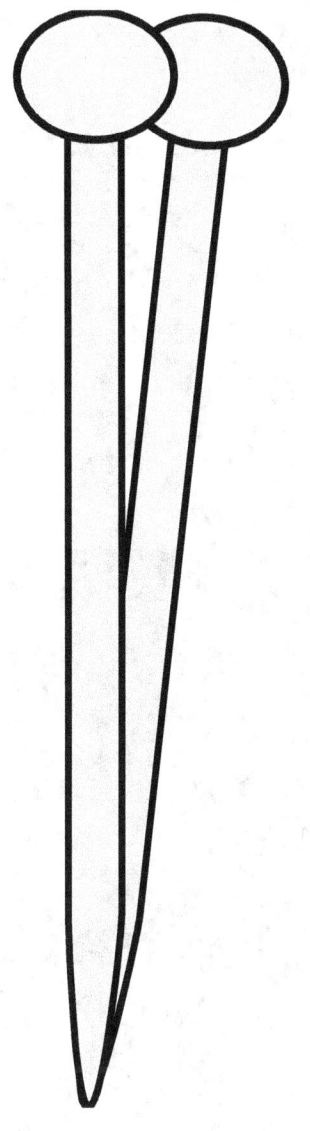

NAILS

5. Saw

The saw is used to cut wood

SAW

6. Wrench

A wrench is used to tighten and loosen the bolts

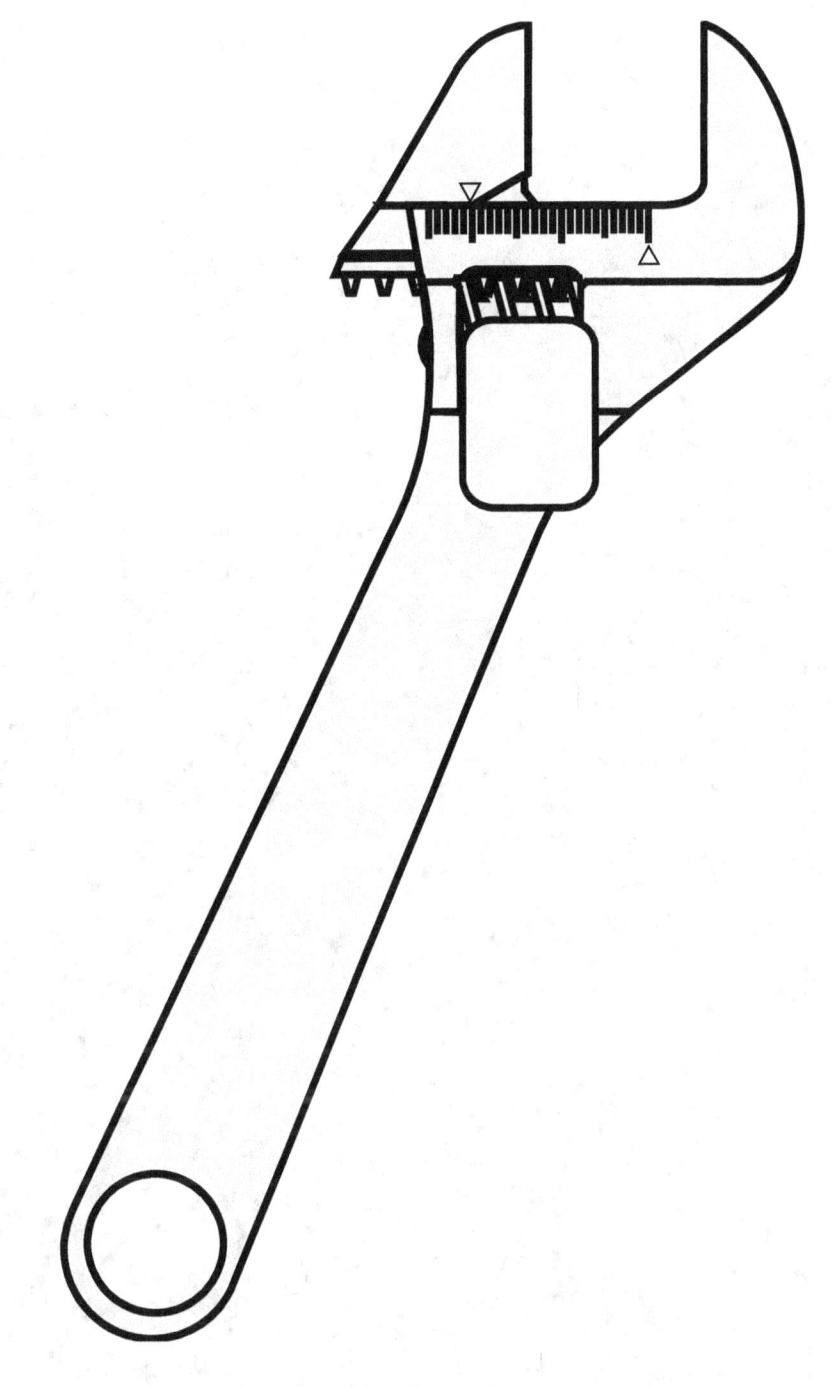

WRENCH

7. Knife

A knife is
used to
cut paper

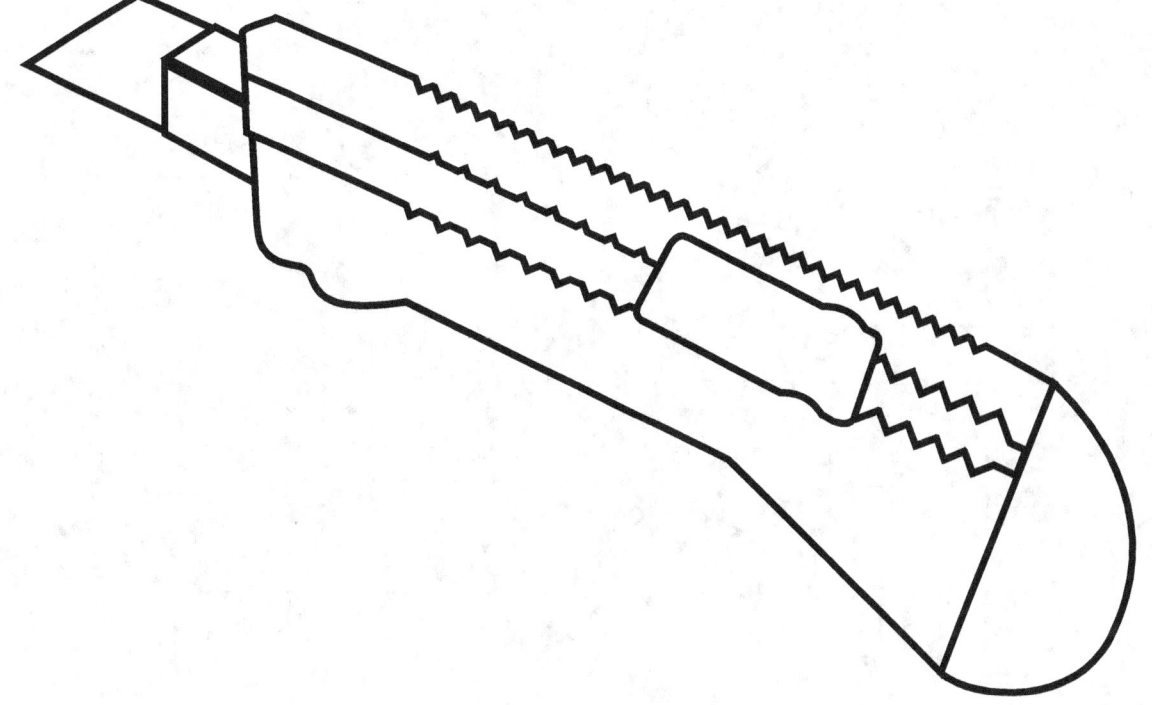

KNIFE

8. Pliers

Pliers are used to bend or cut wires

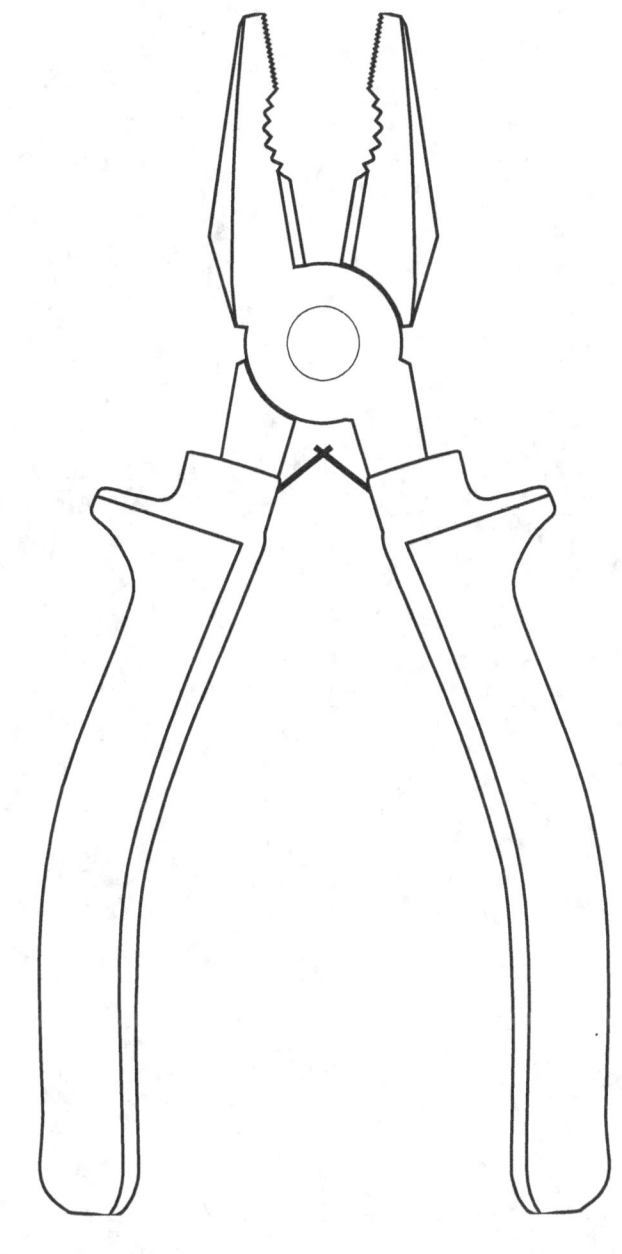

PLIERS

9. Pincers

Pincers are used to remove nails

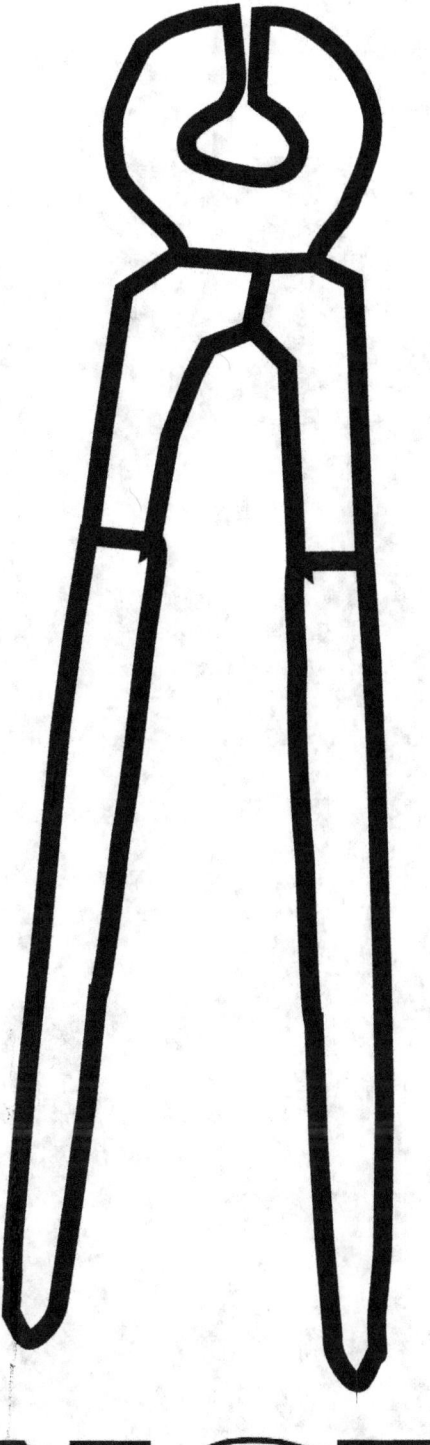

PINCERS

10. Drill

The drill bit is used to drill holes

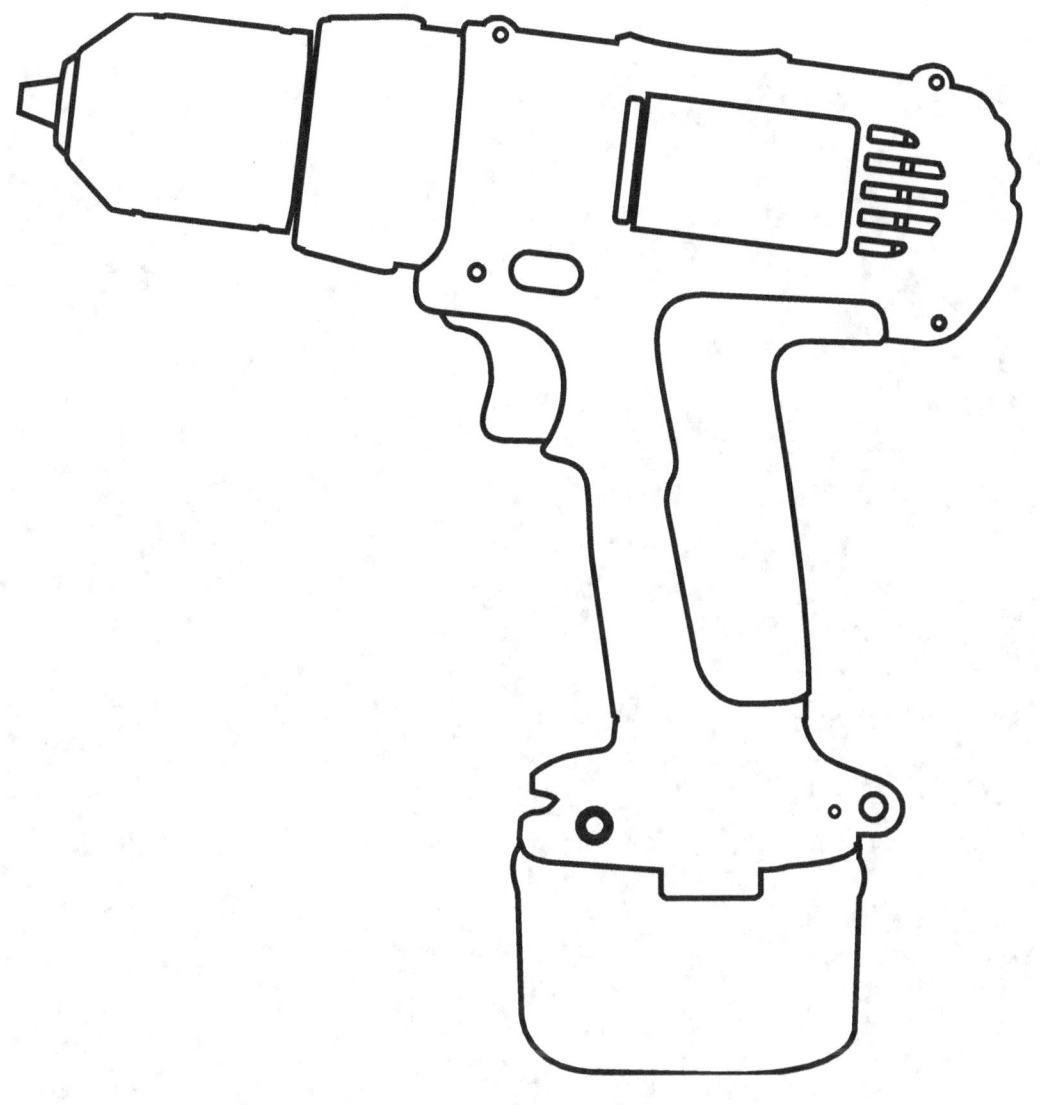

DRILL

11. Spirit level

A spirit level is used to align the horizontal and vertical

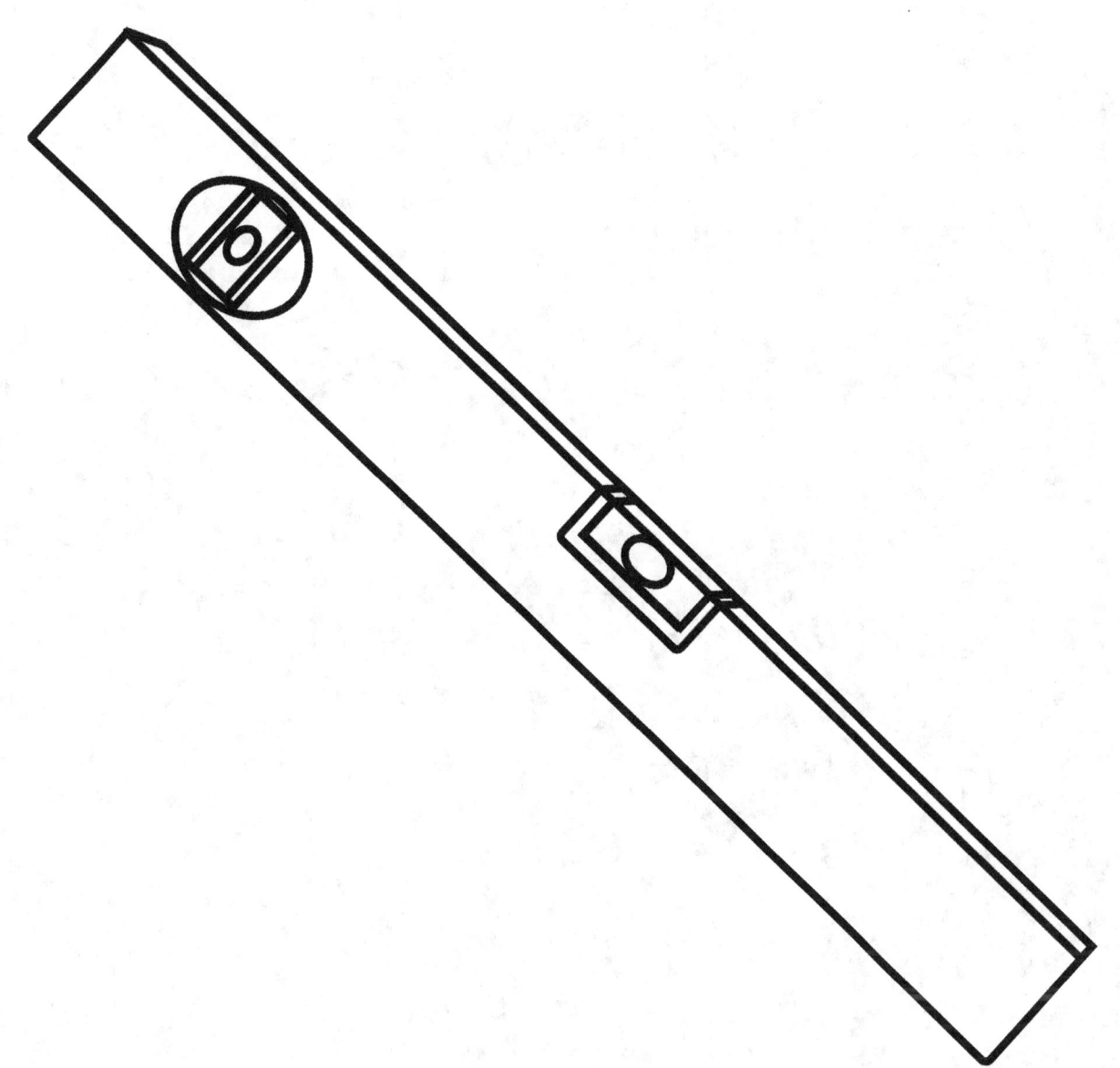

SPIRIT
LEVEL

12. Tape measure

A tape measure is used to measure distances

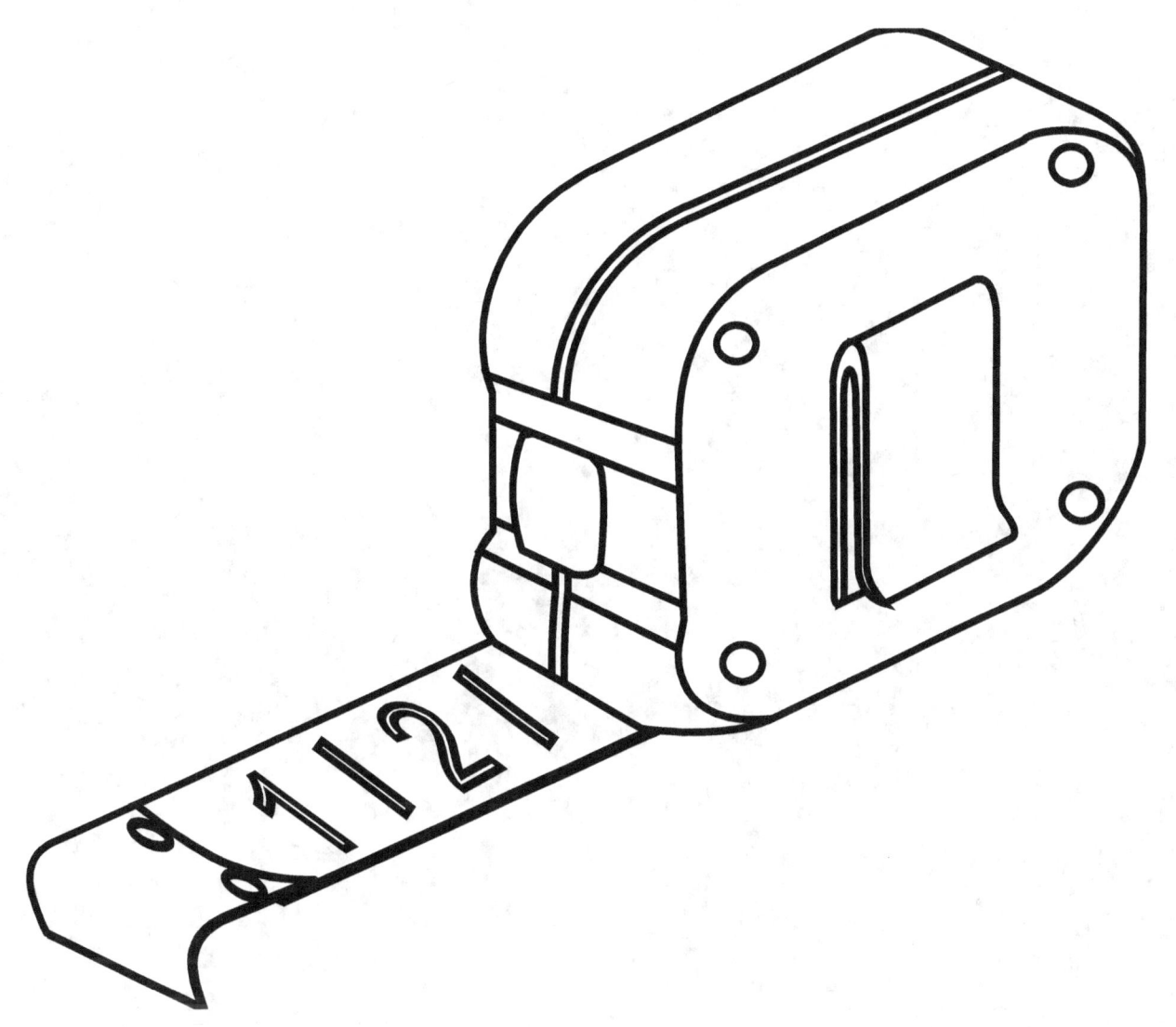

TAPE
MEASURE

13. Spade

The spade is used to dig pits

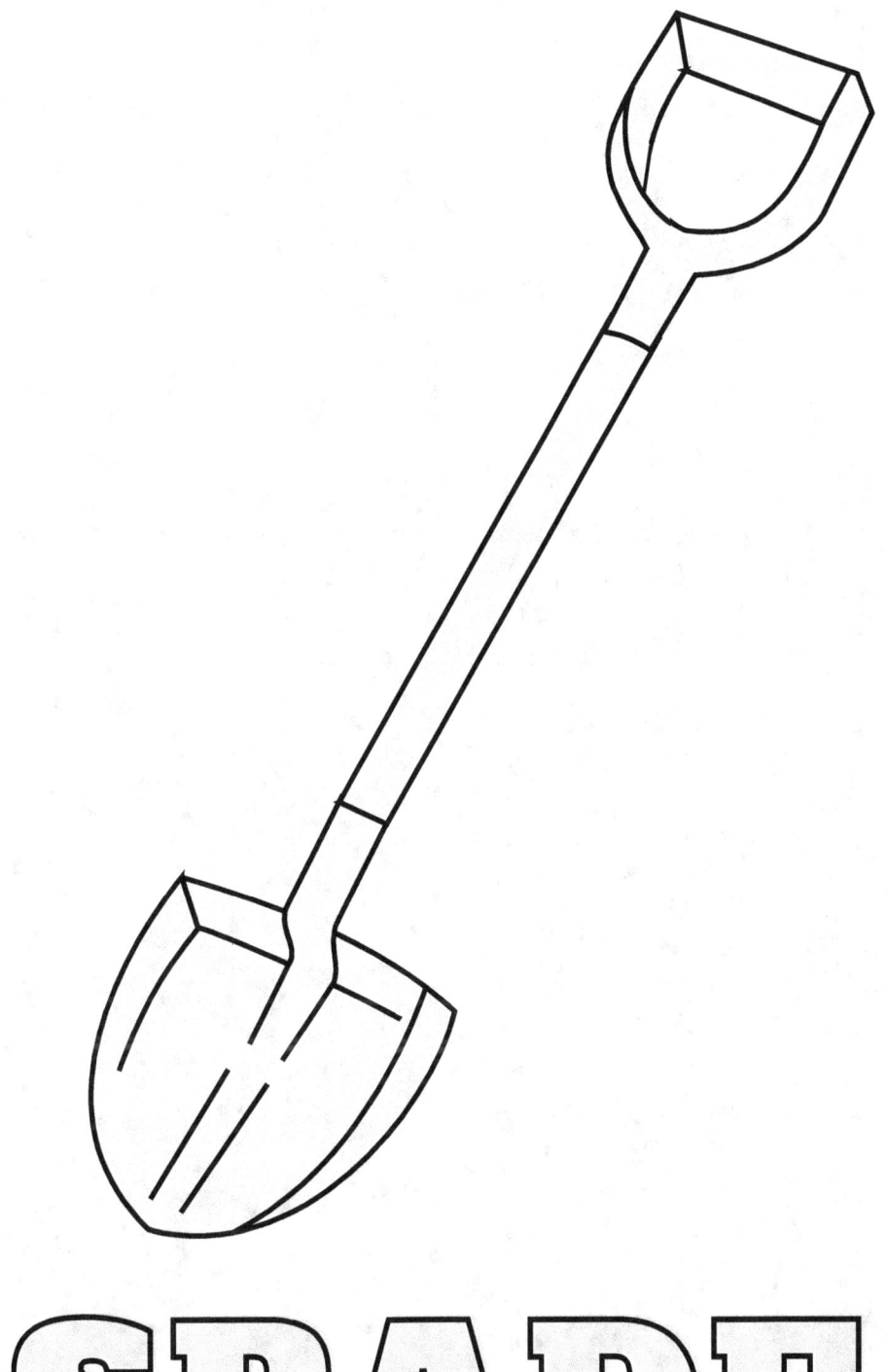

SPADE

14. Spatula

the spatula is used to apply the putty

SPATULA

15. Brush

The brush is used for painting

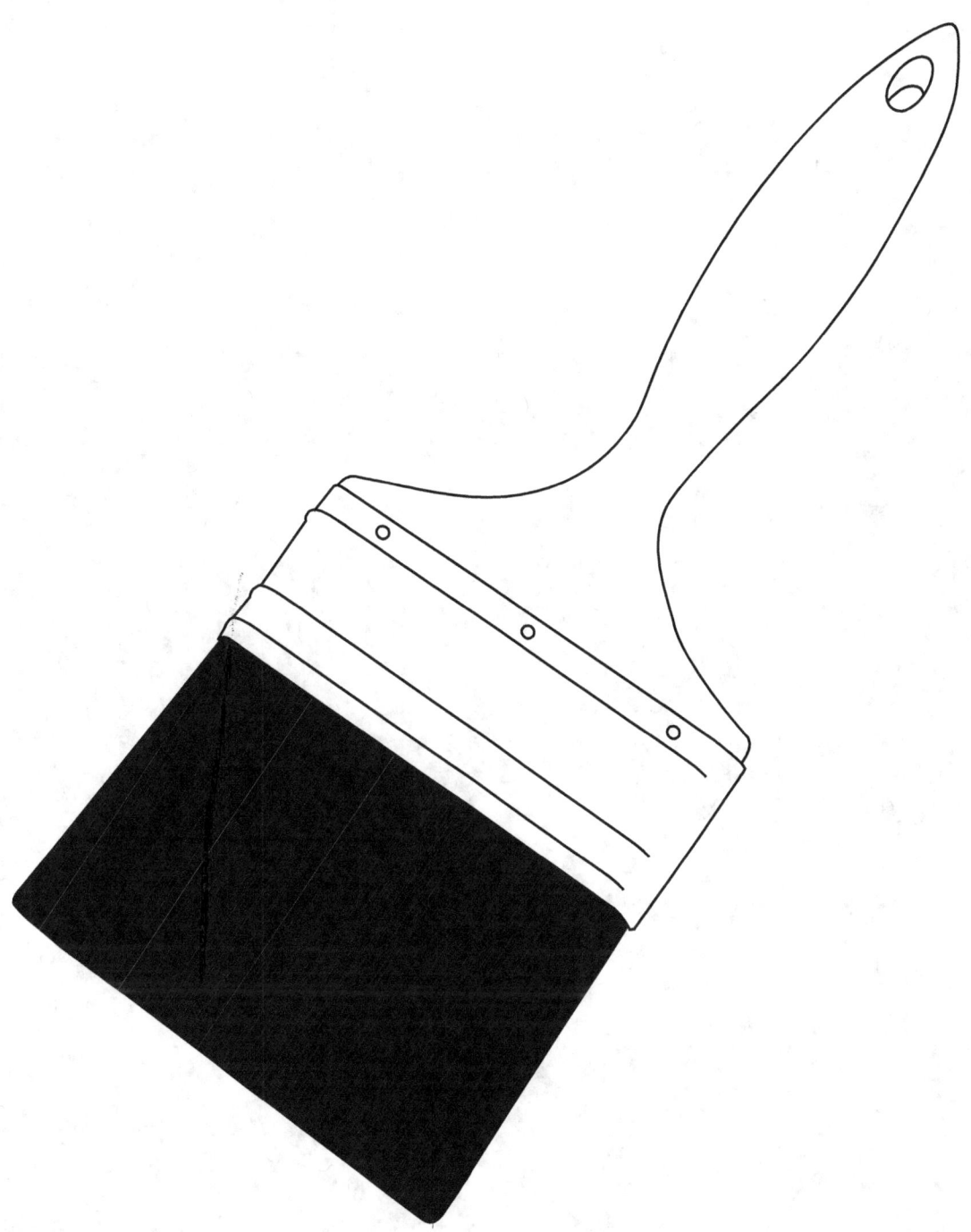

BRUSH

16.Wheel-barrow

The wheelbarrow is used to transport materials

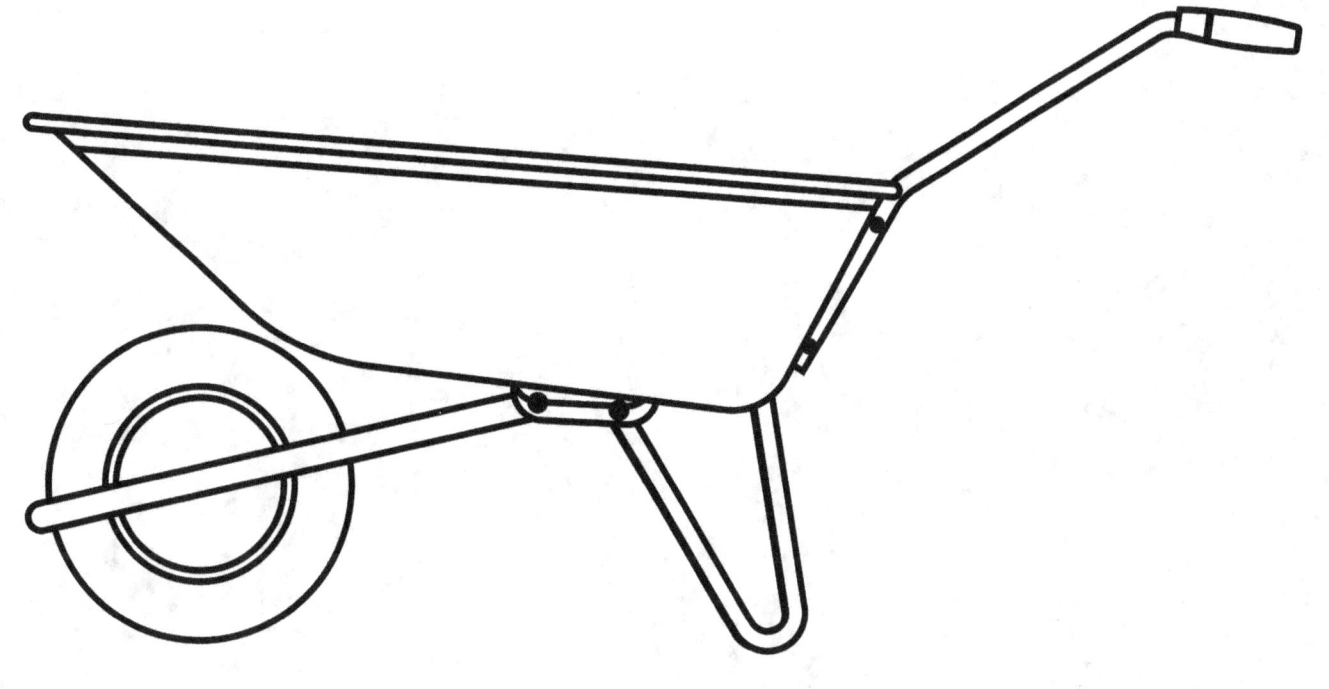

WHELLBARROW

17. Concrete mixer

Concrete mixer is used to mix cement

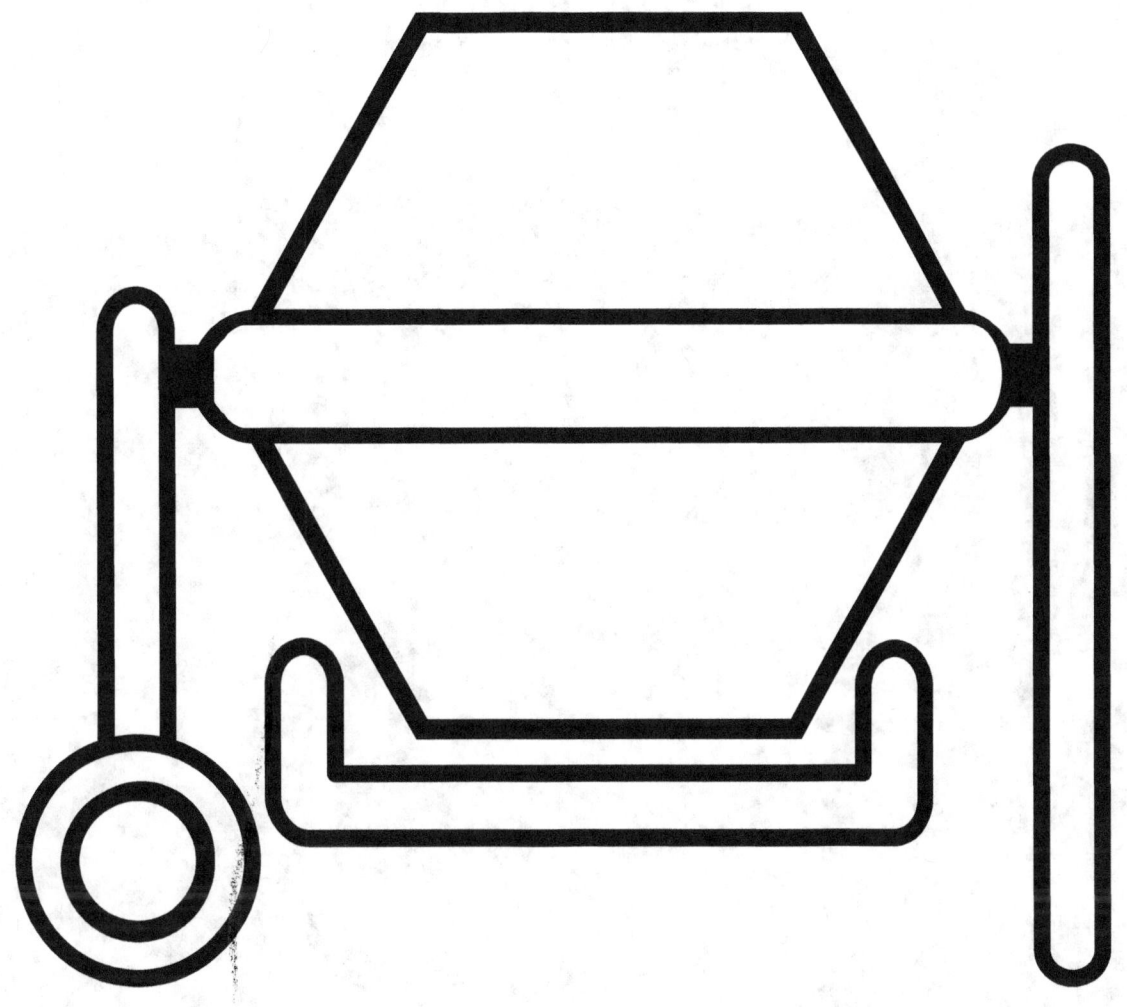

CONCRETE MIXER

18. Generator

The generator produces electricity at the construction site

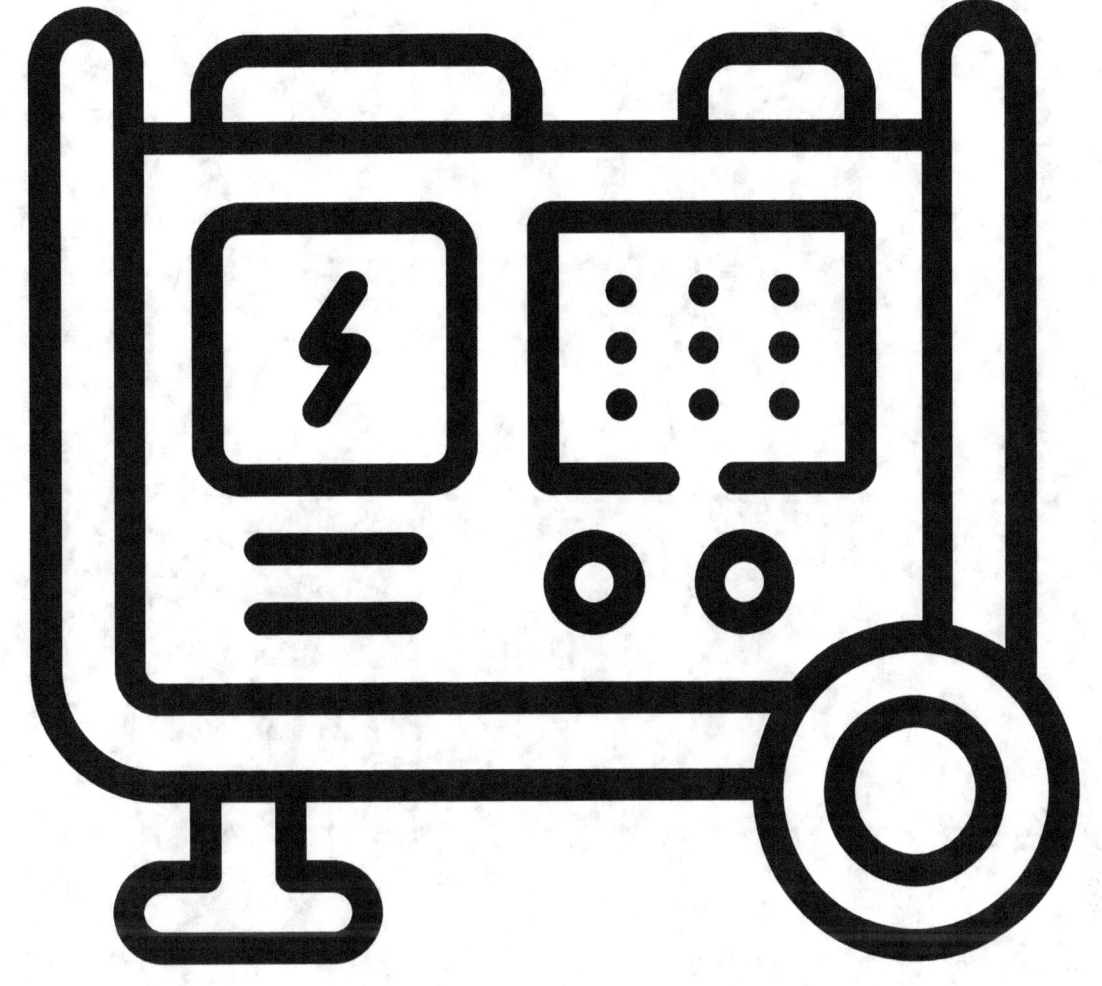

GENERATOR

19. Construction helmet

construction helmet protects the head

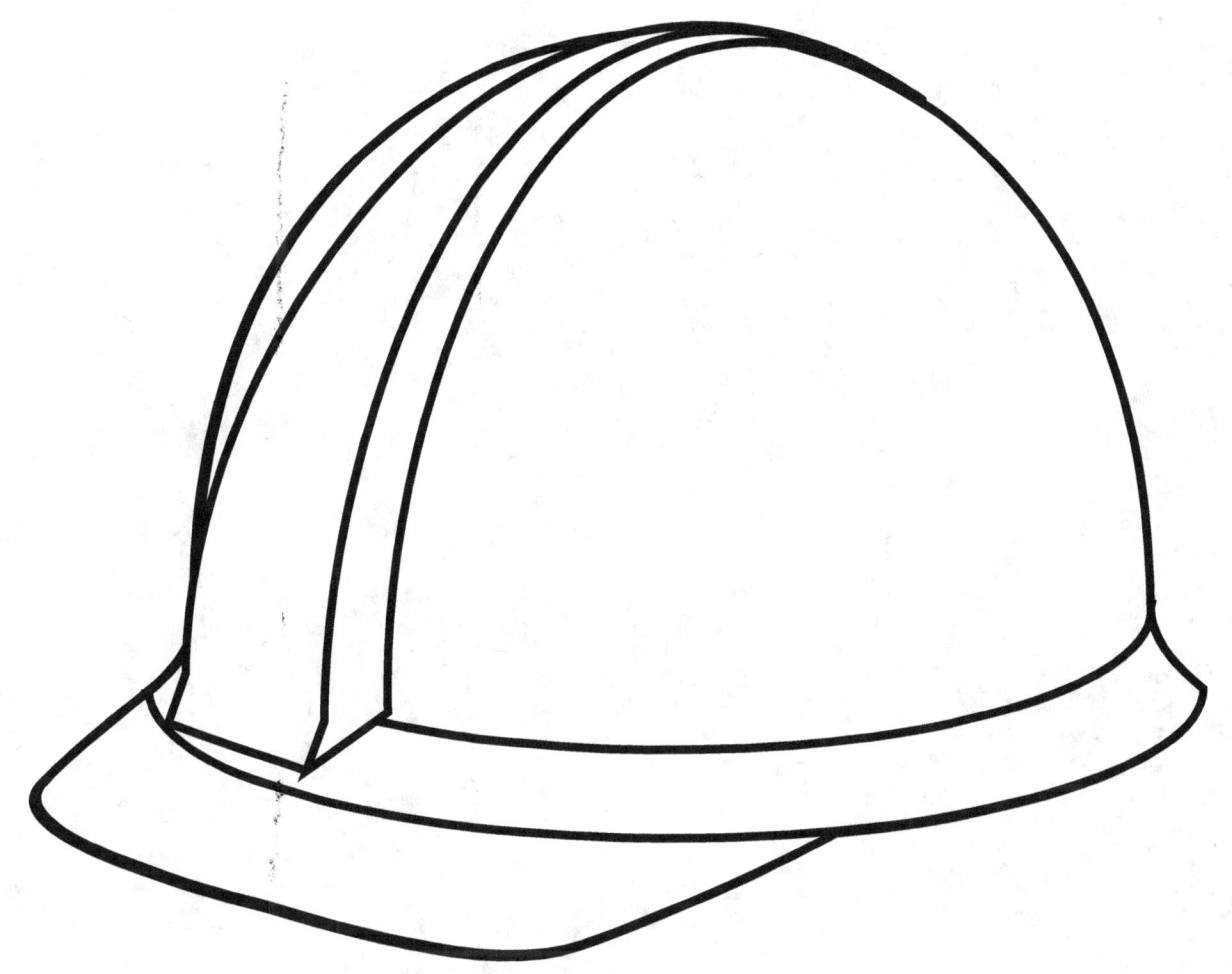

CONSTRUCTION
HELMET

Place for your drawings

Illustration:

pixabay
vecteezy
publicdomainvectors